WORKBOOK/LAB MANUAL
to Accompany

Residential Construction Academy

Carpentry

Second Edition

Floyd Vogt
Kevin Standiford

THOMSON
DELMAR LEARNING

Australia • Brazil • Canada • Mexico • Singapore • Spain • United Kingdom • United States

Workbook/Lab Manual to Accompany
Residential Construction Academy: Carpentry, 2e
Floyd Vogt and Kevin Standiford

Vice President, Technology and Trades ABU:
David Garza

Director of Learning Solutions:
Sandy Clark

Managing Editor:
Larry Main

Acquisitions Editor:
James Devoe

Product Manager:
John Fisher

Marketing Director:
Deborah Yarnell

Marketing Manager:
Kevin Rivenburg

Marketing Coordinator:
Mark Pierro

Director of Production:
Patty Stephan

Production Manager:
Stacy Masucci

Content Project Manager:
Andrea Majot

Technology Project Manager:
Kevin Smith

Editorial Assistant:
Tom Best

Library of Congress Card Catalog Number:
2007003218

ISBN 10: 1-4283-2364-3
ISBN 13: 978-1-4283-2364-3

NOTICE TO THE READER

Publisher does not warrant or guarantee any of the products described herein or perform any independent analysis in connection with any of the product information contained herein. Publisher does not assume, and expressly disclaims, any obligation to obtain and include information other than that provided to it by the manufacturer.

The reader is expressly warned to consider and adopt all safety precautions that might be indicated by the activities herein and to avoid all potential hazards. By following the instructions contained herein, the reader willingly assumes all risks in connection with such instructions.

The publisher makes no representation or warranties of any kind, including but not limited to, the warranties of fitness for particular purpose or merchantability, nor are any such representations implied with respect to the material set forth herein, and the publisher takes no responsibility with respect to such material. The publisher shall not be liable for any special, consequential, or exemplary damages resulting, in whole or part, from the readers' use of, or reliance upon, this material.

Table of Contents

CHAPTER 4 Fasteners

CHAPTER 5 Blueprints, Codes, and Building Layout

CHAPTER **6** Concrete Form Construction 35

CHAPTER **7** Floor Framing 43

CHAPTER **8** Wall and Ceiling Framing 47

CHAPTER **9** Scaffolding, Ladders, and Sawhorses

CHAPTER **10** Roof Framing

CHAPTER **11** **Windows and Doors** 71

CHAPTER **12** **Roofing** 79

CHAPTER **16** Stair Framing and Finish 111

CHAPTER **17** Cabinet and Countertops 119

Preface

Introduction

Designed to accompany *Residential Construction Academy: Carpentry,* second edition, this workbook provides additional review questions and exercises designed to challenge and reinforce the student's comprehension of the content presented in the core text.

About the Text

The workbook is divided into chapters, with each chapter directly corresponding to a chapter in *Residential Construction Academy: Carpentry.* Each chapter consists of an introduction, objectives, review questions, and job sheets and exercises. The review questions are composed of a variety of matching, true/false, multiple-choice, short-answer, and essay-style questions based on the materials presented in the core text and workbook.

Job Sheets

Each job sheet consists of an objective for that job sheet, instructions, and either an activity or checklist. The job sheets range in complexity from entry level to more complex problems that require the student to perform calculations.

Features of This Workbook

- Additional review questions and exercises for *Residential Construction Academy: Carpentry.*
- Job sheets with additional exercises and activities designed to reinforce the material presented in the core text book.

Chapter 1 — Hand Tools

OBJECTIVES

Upon completion of this chapter, you will be able to:

Knowledge-Based

- ⊗ Identify and describe the hand tools a carpenter commonly uses.

Skill-Based

- ⊗ Use hand tools in a safe and appropriate manner.
- ⊗ Maintain hand tools in suitable working condition.

Keywords

Crosscut	Kerfs	Rabbet
Dado	Level	Square
Groove	Plumb	Toe
Heel	Plumb Bob	Whet

Introduction

As in any trade, a good understanding of the tools required to perform the required task is essential to mastering that task. Selecting the correct tool for a particular job is not only critical to completing the task well, but it is also one of the major factors that influence safety. In other words, using the wrong tool for a particular job is one of the leading causes of workplace injuries.

Chapter Review Questions and Exercises

COMPLETION

1. _____ is an L-shaped cutout along the edge or end of lumber.

2. _____ is a cut made across the grain of lumber.

3. _____ is the honing of a tool by rubbing the tool on a flat sharpening stone.

4. _____ is a pointed weight attached to a line for testing plumb.

5. _____ is a tool used to mark a layout and mark angles, particularly 90° angles; a term used to describe when two lines or sides meet at a 90° angle; the amount of roof covering that will cover 100 square feet of roof area.

6. _____ is the back end of objects, such as a handsaw or hand plane.

7. _____ is the width of a cut made with a saw.

8. _____ is when an object is aligned perpendicular to the force of gravity.

9. _____ is a cut, partway through and across the grain of lumber.

10. _____ is when an object is aligned with the force of gravity.

11. _____ is the forward end of tools, such as a hand saw and hand plane.

12. _____ is a cut, partway through and running with the grain of lumber.

Name: _____   Date: _____

Job Sheet 1: Hand Tool Checklist

- Upon completion of this job sheet, you should be able to identify the hand tools commonly used in the carpentry trades.

Tool	Present	Condition
Steel tape measure	☐	_____
Torpedo level	☐	_____
2' to 4' level	☐	_____
Carpenter's square	☐	_____
Plumb bob	☐	_____
Chalk line	☐	_____
Combination square	☐	_____
Crosscut saw	☐	_____
Chisel set ¼" to 1 ½" in width	☐	_____
Rip saw	☐	_____
Combination saw	☐	_____
Hacksaw	☐	_____
Cabinetmaker's chisel	☐	_____
Back saw and miter box	☐	_____
Framing chisel	☐	_____
Keyhole saw	☐	_____
Mortise chisel	☐	_____
Coping saw	☐	_____
Mallet	☐	_____
Claw hammer	☐	_____
Ripping hammer	☐	_____
Tack hammer	☐	_____
Ball peen hammer	☐	_____
Sledge hammer	☐	_____
Nail set	☐	_____
Pry bar	☐	_____
Brads, nails, and spikes	☐	_____
Conventional screwdrivers	☐	_____
Phillips-head screwdrivers	☐	_____
Cordless screwdriver	☐	_____
Wood screws	☐	_____
Sheet metal screws	☐	_____

(continue)

Tool	Present	Condition
Machine screws	☐	_____
Open-end wrench	☐	_____
Box or socket wrench	☐	_____
Allen wrench	☐	_____
Locking wrench	☐	_____
Strap wrench	☐	_____
Pipe wrenches (2)	☐	_____
Lag screws	☐	_____
Bolts	☐	_____
Slip joint pliers	☐	_____
Lineman's pliers	☐	_____
Channel lock pliers	☐	_____
Long-nosed pliers	☐	_____
End-cutting nippers	☐	_____
Polyvinyl	☐	_____
Resorcinol and formaldehyde	☐	_____
Contact cement	☐	_____
Epoxy	☐	_____
C-clamp	☐	_____
Bar clamp	☐	_____
Spring clamp	☐	_____
Hand screw	☐	_____
Vise	☐	_____
Block plane	☐	_____
Trimming plane	☐	_____
Smooth plane	☐	_____
Scrub plane	☐	_____
Jack plane	☐	_____
Fore and joiner planes	☐	_____
Rabbet plane	☐	_____
Grooving plane	☐	_____
Sanding block	☐	_____
Sanding cloth	☐	_____
Sandpapers	☐	_____
Steel wool	☐	_____
Single cut	☐	_____
Double cut	☐	_____
Rasp	☐	_____
Sawhorses	☐	_____
Shop vacuum	☐	_____

Instructor's Response

Name: _____   **Date:** _____

Job Sheet 2: Selecting the Right Hammer

- Upon completion of this job sheet you will be able to select the right hammer for a particular job.
- List the intended use for each hammer listed below

Common Nail Hammer with Curved Claw

_____

_____

Rip Hammer with Straight Claw

_____

_____

Finishing Hammer

_____

_____

Ball Peen Hammer

_____

_____

Hand Drilling Hammer

_____

_____

Soft-Face Hammer

_____

_____

Tack Hammer

_____

_____

Brick Hammer

_____

_____

Drywall (Wallboard) Hammer

_____

_____

Carpenter's Mallet

_____

_____

Instructor's Response

Chapter 2   Power Tools

OBJECTIVES

Upon completion of this chapter, you will be able to:

Knowledge-Based

- State general safety rules for operating power tools.
- Describe and safely use the following: circular saws, saber saws, reciprocating saws, drills, hammer-drills, screwdrivers, planes, routers, sanders, staplers, nailers, and power-actuated drivers.
- Describe and adjust the table saw and power miter saw.

Skill-Based

- Safely crosscut lumber to length, rip to width, and make miters by using a table saw.
- Safely crosscut to length, making square and miter cuts by using a power miter saw.

Keywords

Bevel	Crosscut	Miter Gauge
Chamfer	Fence	Rip
Compound	Miter	

Introduction

In the previous chapter, the topic of hand tools was introduced. It was stated that a good working knowledge of the hand tools used in the trade will not only help the carpenter in doing a better job, but it will also help prevent accidents. The same is true for power tools. However, with power tools the need for safety is ever more important. Injuries can be more severe than with hand tools.

Chapter Review Questions and Exercises

COMPLETION

1. _____ is a guide used on a table saw for making miters and square ends.
2. _____ is sawing lumber in the direction of the grain.
3. _____ is the sloping edge or side of a piece at any angle other than a right angle.

4. _____ is an edge or end bevel that does not go all the way across the edge or end.

5. _____ is a cut made across the grain of lumber.

6. _____ is a bevel cut across the width and also through the thickness of a piece of wood.

7. _____ is a guide for ripping lumber on a table saw.

8. _____ is the cutting of the end of a piece of wood at any angle other than a right angle.

Name: _____   **Date:** _____

Job Sheet 1: Power Tool Checklist

- Upon completion of this job sheet, you should be able to identify the power tools commonly used in the carpentry trades.

Tool	Present	Condition
Circular saw	☐	_____
Band saw	☐	_____
Jig saw	☐	_____
$\frac{3}{8}''$ variable-speed drill	☐	_____
Reciprocating saw	☐	_____
Table saw (Bench saw)	☐	_____
Cordless drill	☐	_____
Router	☐	_____
Radial-arm saw	☐	_____
Orbital sander	☐	_____
Belt sander	☐	_____
Disk sander	☐	_____

Instructor's Response

Chapter 3

Wood and Wood Products

OBJECTIVES

Upon completion of this chapter, you will be able to:

Knowledge-Based

- Define hardwood and softwood and give examples of some common kinds.
- State the grades and sizes of lumber.
- Describe the composition, kinds, sizes, and several uses of: plywood, oriented strand board, particleboard, hardboard, medium-density fiberboard, and soft board.
- Describe the uses and sizes of: laminated veneer lumber, parallel strand lumber, laminated strand lumber, wood I-beams, and glue-laminated beams.

Skill-Based

- Calculate linear feet and compute square foot and board foot measure.

Keywords

Air Dried	Dry Kiln	Panel
Annular Rings	Finger Joint	Pith
Board	Header	Plain-Sawed
Board Foot	Heartwood	Quarter-Sawed
Cambium Layer	Lumber	Sapwood
Coniferous	Medullar Rays	Sawyer
Deciduous	Millwork	Tempered
Dimension	On Center (OC)	Timbers

Introduction

The primary material used in residential construction is lumber. Lumber is a product made from forest products. It is manufactured from a variety of different specimens and constructed into a wide range of different shapes and sizes. In the past when lumber was produced, there was an abundance of waste. However, as an effort to reduce the amount of waste and improve lumber strength, the industry of engineered lumber has emerged.

As stated earlier, lumber comes in a variety of different sizes; however, the sizes that are typically used when lumber is specified and purchased is its nominal size. The nominal size of a board varies from the actual size of the board. This is due to planing, and shrinkage as the board is dried. This results in the final lumber being slightly smaller than the nominal size. The following charts show the nominal size and actual size of lumber commonly used in framing.

General Properties of Hardwood and Softwood

The carpenter works with wood more than any other material and must understand its characteristics in order to use it intelligently. Wood is a remarkable substance and is classified as either hardwood or softwood.

There are different methods of classifying these woods. The most common method of classifying wood is by its source.

Hardwood—comes from deciduous trees that shed their leaves each year.

Softwood—comes from coniferous, or cone-bearing trees, commonly know as evergreens.

Common Hardwoods

- Ash
- Birch
- Cherry
- Hickory
- Maple
- Mahogany
- Oak
- Walnut

Common Softwoods

- Pine
- Fir
- Hemlock
- Spruce
- Cedar
- Cypress
- Redwood

The best way to learn the different kinds of woods is by doing the following:

- Look at the color and the grain.
- Feel if it is heavy or light.
- Feel if it is hard or soft.
- Smell it for a characteristic odor.

Size Chart

Framing Lumber

Nominal Size	Actual Size
2 × 2	1½ × 1½
2 × 3	1½ × 2½
2 × 4	1½ × 3½
2 × 6	1½ × 5½
2 × 8	1½ × 7¼
2 × 10	1½ × 9¼
2 × 12	1½ × 11¼
4 × 4	3½ × 3½
4 × 6	3½ × 5½
4 × 10	3½ × 9¼
6 × 6	5½ × 5½

Boards

Nominal Size	Actual Size
1 × 2	¾ × 1½
1 × 3	¾ × 2½
1 × 4	¾ × 3½
1 × 5	¾ × 4½
1 × 6	¾ × 5½
1 × 8	¾ × 7¼
1 × 10	¾ × 9¼
1 × 12	¾ × 11¼

Chapter Review Questions and Exercises

COMPLETION

1. _____ is a technique of removing water from lumber by using natural wind currents.

2. _____ are the rings seen when viewing a cross-section of a tree trunk; each ring constitutes one year of tree growth.

3. _____ is a layer just inside the bark of a tree where new cells are formed.

4. _____ is the wood in the inner part of a tree, usually darker and containing inactive cells.

5. _____ is a term used to define a measurement of an item; also used to refer to all 2× lumber used in framing.

6. _____ is a general term for wood that is cut from a log to form boards, planks, and timbers.

7. _____ is a large oven used to remove water from lumber.

8. _____ is lumber usually less than 2 inches thick.

9. _____ is a measure of lumber volume that equals 1 foot square and 1 inch thick or any equivalent lumber volume. The letter M is used to represent 1,000 board feet.

10. _____ are bands of cells radiating from the cambium layer to the pith of a tree to transport nourishment toward the center.

11. _____ is any wood product that has been manufactured, such as moldings, doors, windows, and stairs for use in building construction; sometimes called joinery.

12. _____ is a method of sawing lumber that produces flat-grain where annular rings tend to be parallel to the width of the board.

13. _____ is the distance from the center of one structural member to the center of the next one.

14. _____ are trees that are cone-bearing; also known as evergreen trees.

15. _____ is the outer part of a tree just beneath the bark containing active cells.

16. _____ is a person whose job is to cut logs into lumber.

17. _____ is a treatment process in which the material is made harder and stronger.

18. _____ are large pieces of lumber over 5 inches in thickness and width.

19. _____ are trees that shed leaves each year.

20. _____ is the process in which shorter lengths are glued together using deep, thin V grooves, resulting in longer lengths.

21. _____ is framing members placed at right angles to joists, studs, and rafters to form and support openings.

22. _____ is a large sheet of building material that usually measures 4 × 8 feet.

23. _____ is a method of sawing lumber that produces a close grain pattern where the annular rings tend to be perpendicular to the width of the board.

24. _____ is the small, soft core at the center of a tree.

Name: _____ Date: _____

Job Sheet 1: Calculating Board Feet

- Upon completion of this job sheet, you will be able to calculate the number of board feet.

Calculate the board feet of the lumber listed below.

Number of Pieces	Length	Width	Thickness	Number of Board Feet
23	8	6	1	
12	10	8	2	
45	8	4	2	
5	12	10	1	
120	8	2	2	
37	10	10	2	
56	12	12	1	
12	8	6	2	
25	8	4	1	

Instructor's Response

Chapter 4   Fasteners

OBJECTIVES

Upon completion of this chapter, you will be able to:

Knowledge-Based

⊘ Name and describe the following commonly used fasteners and select them for appropriate use:

⊘ nails

⊘ screws

⊘ lag screws

⊘ bolts

⊘ solid wall anchors

⊘ hollow wall anchors

⊘ adhesives

Keywords

Anchor	Face Nail	Penny (d)
Box Nail	Finish Nail	Toenail
Duplex Nail	Galvanized	
Electrolysis	Mastic	

Introduction

Just as important as lumber is the way in which it is fastened and anchored. A *fastener* is a mechanical device that is used to mechanically join two or more mating surfaces or objects. Fasteners are now on the market for just about any job. They can be used where wood meets wood, concrete, or brick, and most are approved by the Uniform Building Code requirements. However, you should always consult your local building code before selecting a particular type of fastener to incorporate.

Nails

The following is a list of the various types of nails:

- **Common nail**—most often used of all nails. Used for most applications where special features of other nail types are not needed.

- **Box nails**—used for boxes and crates.

- **Finishing nails**—can be driven below the surface of the wood and concealed with putty so that they are completely hidden.

- **Casing nails**—used for installing exterior doors and windows.
- **Duplex nails**—used for temporary structures, such as locally built scaffolds.
- **Roofing nails**—used for installing asphalt and fiberglass roofing shingles.
- **Masonry nails**—used when nailing into concrete or masonry.

Screws

Screws are used when stronger joining power is needed, or for when other materials must be fastened to wood. The screw is tapered to help draw the wood together as the screw is inserted. Screw heads are usually flat, oval, or round, and each has a specific purpose for final seating and appearance.

Types of Screws

- **Drywall screws**—used to attach drywall to wall studs.
- **Sheet metal screws**—used to fasten metal to wood, metal, plastic, or other materials. Sheet metal screws are threaded completely from the point to the head, and the threads are sharper than those of wood screws. Machine screws are for joining metal parts, such as hinges to metal door jambs.
- **Particleboard and deck screws**—corrosion-resistant screws used for installing deck materials and particleboard.
- **Lag screws**—used for heavy holding and are driven in with a wrench rather than a screwdriver.

Tip: Screw length should penetrate two-thirds of the combined thickness of the materials being joined. Use galvanized or other rust-resistant screws where rust could be a problem.

Screw head shapes are usually determined by the screw types and the pitch and depth of the threads. The most common are oval, pan, bugle, flat, round, and hex. There are also different types of slots for these screws.

Bolts

Nuts and bolts are usually used at the same time: The bolt is inserted through a hole drilled in each item to be fastened together, and then the nut is threaded onto the bolt from the other side and tightened, to give a strong connection. Using nuts and bolts also allows for the disassembly of parts.

Types of Bolts

- **Cap screws**—available with hex heads, slotted heads, Phillips head, and Allen drive.
- **Stove bolts**—either round or flat heads and threaded all the way to the head; used to join sheet metal parts.
- **Carriage bolts**—a round-headed bolt for timber; threaded along part of the shank; inserted into holes already drilled.

Adhesives

Sometimes adhesives are needed when nails or screws alone cannot hold materials together.

Types of Adhesives

Carpenter's wood glue—a white, creamy glue usually available in convenient plastic bottles. Mainly used for furniture, craft, or woodworking projects, polyvinyl carpenter's wood glue sets in an hour, dries clear, and won't stain. However, it is vulnerable to moisture.

Epoxy—the only adhesive with a strength greater than the material it bonds. It resists almost anything, from water to solvents. Epoxy can be used to fill cavities that would otherwise be difficult to bond. Use it in warm temperatures but read the manufacturer's instructions carefully, since drying times vary and mixing the resin and hardener must be exact.

Tips for using epoxy:

- One gallon of epoxy will cover: 12.8 square feet at ⅛-inch thickness, 6.4 square feet at ¼ inch thickness.

- Although epoxy generally is able to withstand high temperatures for short periods, we do not recommend using it in conditions above 200°F.

- To achieve maximum adhesion, remove oil, dirt, rust, paint, and water. Use a degreasing solvent (alcohol or acetone) to remove oil and grease. Sand or wipe away paint, dirt, or rust. Roughing up the surface increases surface area for a better bond.

- Epoxy cures quickly enough that there is significant strength after 3 hours. At 70°F the working time is 15 minutes; however, it is still possible to reposition work after up to 45 minutes.

- Uncured epoxy will clean up with soap and water, or denatured alcohol. Wash contaminated clothing. Cured epoxy can be removed by scraping, cutting, or removing in layers with a good paint remover.

Contact Cements

Contact cement is used to bond veneers or to bond plastic laminates to wood for table tops and counters. Coat both surfaces thinly and allow to dry somewhat before bonding. Align the surfaces perfectly before pressing together, these adhesives do not allow repositioning. Use in a well-ventilated area.

Types of Fasteners

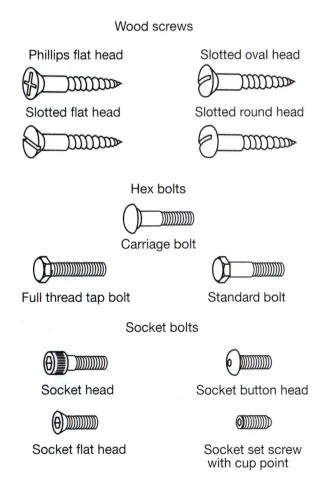

Wood screws

Phillips flat head

Slotted oval head

Slotted flat head

Slotted round head

Hex bolts

Carriage bolt

Full thread tap bolt

Standard bolt

Socket bolts

Socket head

Socket button head

Socket flat head

Socket set screw with cup point

Washers

Flat washer
(USS and SAE)

Lock washer
internal tooth

Lock washer

Finishing washer

Lock washer
external tooth

Dock washer

Sheet metal screws

Phillips flat head

Phillips pan head self drilling

Phillips oval head

Hex washer head

Phillips truss head

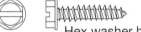

Hex washer head self drilling

Phillips pan head

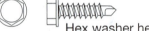

Hex washer head self drilling
with sealing washer

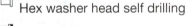

Machine screws

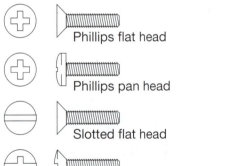

Phillips flat head

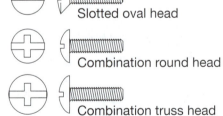

Slotted oval head

Phillips pan head

Combination round head

Slotted flat head

Combination truss head

Phillips oval head

Slotted round head

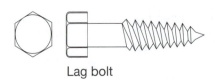

Lag bolt

Nuts

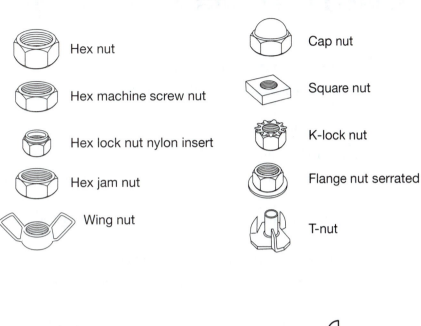

Hex nut

Cap nut

Hex machine screw nut

Square nut

Hex lock nut nylon insert

K-lock nut

Hex jam nut

Flange nut serrated

Wing nut

T-nut

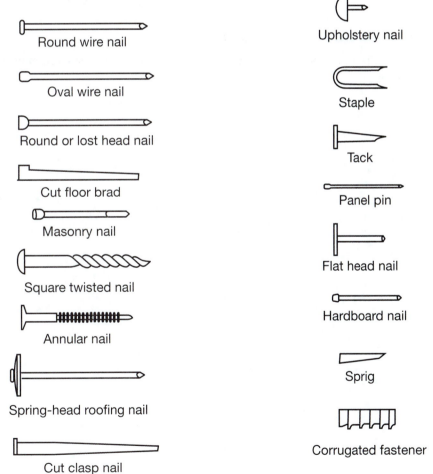

Round wire nail

Upholstery nail

Oval wire nail

Staple

Round or lost head nail

Tack

Cut floor brad

Panel pin

Masonry nail

Flat head nail

Square twisted nail

Hardboard nail

Annular nail

Sprig

Spring-head roofing nail

Corrugated fastener

Cut clasp nail

Chapter Review Questions and Exercises

COMPLETION

1. _____ is a device used to fasten structural members in place.

2. _____ is the method of driving a nail straight through a surface material into a supporting member.

3. _____ is a thick adhesive.

4. _____ is a thin, short, finishing nail.

5. _____ is a thin nail with a head, usually coated with a material to increase its holding power.

6. _____ is a double-headed nail used for temporary fastening such as in the construction of wood scaffolds.

7. _____ is protected from rusting by a coating of zinc.

8. _____ is the decomposition of one of two unlike metals in contact with each other in the presence of water.

9. _____ is a thin nail with a small head designed for setting below the surface of finish material.

10. _____ is a term used in designating nail sizes.

11. _____ is a method of driving a nail diagonally through a surface material into a supporting member.

Name: _____   **Date:** _____

Job Sheet 1: Selecting the Correct Nail

- Upon completion of this job sheet, you will be able to select the correct nail for a particular purpose.

Nail Type	Description/Use
Common	
Box	
Finishing	
Casting	
Duplex	
Roofing	
Masonry	

Instructor's Response

Name: _____   Date: _____

Job Sheet 2: Selecting the Correct Screw

• Upon completion of this job sheet, you will be able to select the correct screw for a particular purpose.

Types of Screws	Description/Use
Drywall	
Sheet metal	
Particleboard and deck	
Lag	

Instructor's Response

Chapter 5 — Blueprints, Codes, and Building Layout

OBJECTIVES

Upon completion of this chapter, you will be able to:

Knowledge-Based

- Describe and explain the function of the various kinds of drawings contained in a set of blueprints.
- Demonstrate how specifications are used.
- Identify various types of lines and read dimensions.
- Define and explain the purpose of building codes and zoning laws.
- Explain the requirements for obtaining a building permit and the duties of a building inspector.
- Identify and explain the meaning of symbols and abbreviations used on a set of prints.

Skill-Based

- Establish level points across a building area by using a water level and by using a carpenter's hand spirit level in combination with a straight edge.
- Accurately set up and use the builder's level, transit level, and laser level.
- Use an optical level to determine elevations.
- Lay out building lines by using the Pythagorean theorem and check the layout for accuracy.
- Build batter boards and accurately establish building lines with string.
- Read and interpret plot, foundation, floor, and framing plans.

Keywords

Detail	Laser	Pythagorean Theorem
Elevation	Ledger	Section
Foundation	Plan	

Introduction

A building code is a set of rules that specify the minimum acceptable level of safety for constructed objects such as buildings and non-building structures.

The International Building Code (IBC) is a combination of three model building codes—the Uniform Building Code (UBC), the National Building Code (BOCA), and the Standard Building Code (SBC). The IBC has a great influence on every aspect of construction management. It is divided into 35 chapters and 11 appendices.

The codes can be used effectively during the initial planning phase as well as when designing a building. When the construction is still in the initial design phase, you must determine the following five basic classifications of your building:

- Occupancy group—It describes what the structure is going to be used for (warehouse, theater, school, and so on). A single structure can house multiple occupancy groups.

- Location on the site—The physical placement of the structure on the property, particularly with respect to the property lines and other structures.

- Floor area—The amount of space on each of the building's floors or stories.

- Height or number of stories—The height of the building in feet above the grade plane (a representative level of the ground around the building); the number of floors in the structure above the grade plane.

- Construction type—A way of classifying the construction of a building based upon the types of materials used to construct the structure and the fire rating of the building's structure.

The Americans with Disabilities Act

The innovations in medical and scientific technology have enabled people to survive conditions that were previously considered fatal. The Americans with Disabilities Act (ADA) strives toward giving citizens with disabilities a better chance to live productive and fulfilling lives. The ADA is a federal law requiring all new public accommodations and commercial facilities to be accessible and usable by people with disabilities.

Most of the requirements are meant to accommodate people in wheelchairs, but the ADA also addresses the problems of the elderly, the blind, the hearing impaired, and those with other forms of physical disabilities.

The primary issues addressed by the ADA are access to buildings (stairs, elevators), the building spaces they contain (parking requirements), and the use of facilities such as rest rooms, drinking fountains, and telephones.

Chapter Review Questions and Exercises

COMPLETION

1. _____ in an architectural drawing, an object drawn as viewed from above.

2. _____ is a close-up view of a plan or section.

3. _____ is that part of a wall on which the major portion of the structure is erected.

4. _____ is a temporary or permanent supporting member for joists or other members running at right angles; horizontal member of a set of batter boards.

5. _____ is a drawing in which the height of the structure or object is shown; also, the height of a specific point in relation to another reference point.

6. _____ is a concentrated, narrow beam of light; optical leveling and plumbing instrument used in building construction.

7. A _____ is a drawing showing a vertical cut-view through an object or part of an object.

8. _____ is a mathematical expression that states the sum of the square of the two sides of a right triangle equals the square of the diagonal side.

Name: _____   **Date:** _____

Job Sheet 1: Identifying Abbreviations

- Upon completion of this job sheet, you will be able to identify common abbreviations used on drawings.

Abbreviation	Description
ABS	
ADA	
BS	
DWV	
DH	
FU	
O_2	
PSI	
PRV	
UNO	

Instructor's Response

Name: _____   Date: _____

Job Sheet 2: Identifying Line Types

- Upon completion of this job sheet, you will be able to identify common line types used on drawings.

A. –   _____

B. — – — – — – — – — – — – —   _____

C. _____   _____

D. — – – — – – — – – — – – —   _____

E. — — — — — — — — — —   _____

F. — – – — – – — – – — – – —   _____

Instructor's Response

Chapter 6

Concrete Form Construction

OBJECTIVES

Upon completion of this chapter, you will be able to:

Knowledge-Based

- ○ Explain techniques used for the proper placement and curing of concrete.
- ○ Describe the composition of concrete and factors affecting its strength, durability, and workability.
- ○ Explain the reasons for making a slump test.
- ○ Explain the reasons for reinforcing concrete and describe the materials used.

Skill-Based

- ○ Estimate quantities of concrete.
- ○ Construct forms for footings, slabs, walks, and driveways.
- ○ Construct concrete forms for foundation walls.
- ○ Lay out and build concrete forms for stairs.

Keywords

Buck	Gusset	Run
Concrete	Pilaster	Scab
Concrete Block	Portland Cement	Spreader
Frost Line	Reinforcing Rods	Stud
Girder	Rise	Vapor Retarder

Introduction

Concrete is a versatile building material that is used for a variety of different applications (floors, foundations, walls, ceilings, and so on). In commercial construction, it is one of the materials of choice; in residential construction, it has gained popularity and is being used more.

Estimating the Number of Concrete Blocks

To estimate the number of concrete blocks needed to construct a wall, calculate by using the following steps:

1. Calculate the length of the wall: Multiply the length of the wall by 12. (This should be done if the wall dimensions are given in both feet and inches. If the wall dimensions are given in inches only, then the length of the wall does not need to be multiplied by 12.)

2. Calculate the height of the wall: Multiply the height of the wall by 12. (This should be done if the wall dimensions are given in both feet and inches. If the wall dimensions are given in inches only, then the length of the wall does not need to be multiplied by 12.)

3. Calculate the number of blocks needed in a course: Divide the length of the wall calculated in step 1 by the length of the concrete block. For an 8 × 8 × 16 block, this would be: Length of wall/16.

4. Calculate the number of courses needed: Divide the height of the wall calculated in step 2 by the height of the concrete block. For an 8 × 8 × 16 block, this would be: Height of wall/12.

5. Calculate the number of blocks needed: Multiply the number of blocks calculated in step 3 by the number of rows calculated in step 4.

For example, given a wall 20 feet 6 inches (long) × 12 feet (tall), calculate the number of 8 × 8 × 16 with a ⅜-inch mortar joint needed. See picture below:

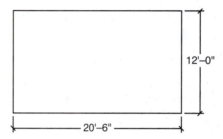

Calculate the length of the wall: 20 × 12 + 6 = 248
Calculate the height of the wall: 12 × 12 = 144
Calculate the block needed in a course: 248/16 = 15.5 blocks
Calculate the number of courses needed: 144/8 = 18
Calculate the number of blocks needed: 15.5 × 18 = 279 blocks

Estimating the Amount of Concrete

To estimate the amount of concrete needed for a particular project, the following steps should performed. See picture below:

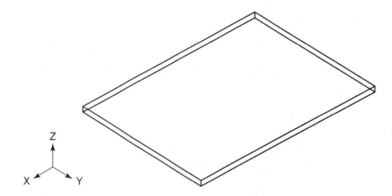

1. Convert the length to feet. For example, 12 feet 6 inches would be 12.5 feet.

2. Convert the width to feet. For example, 8 feet 8 inches would be 8 feet + (8 inches/12) = 8.667 feet.

3. Convert the thickness to feet. For example, 4 inches would be 4/12 = 0.3333 feet.

4. Multiply the length from step 1 by the width from step 2 by the thickness from step 3. For example, 12.5 feet × 8.667 feet × 0.3333 feet = 36.1088 cubic feet.

5. Divide the cubic feet of concrete calculated in step 4 by 27 cubic feet/yard. This will convert the amount of concrete from cubic feet to cubic yards (the unit in which concrete is typically sold). For example, 36.1088 cubic feet/27 cubic feet per yard = 1.3373 cubic yards.

Chapter Review Questions and Exercises

COMPLETION

1. _____ is also called rebar, which are steel bars placed in concrete to increase tensile strength.

2. _____ is a fine gray powder that when mixed with water, forms a paste that sets rock hard; an ingredient in concrete.

3. _____ is a vertical framing member in a wall running between plates.

4. _____ is the vertical distance of the flight; in roofs, the vertical distance from plate to ridge; may also be the vertical distance through which anything rises.

5. _____ is the horizontal distance over which rafters, stairs, and other like members travel.

6. _____ is a concrete masonry unit (CMU) used to make building foundations, typically measuring 8 × 8 × 16 inches.

7. _____ is a block of wood or metal used over a joint to stiffen and strengthen it.

8. _____ is a building material made from Portland cement, aggregates, and water.

9. _____ is a column built within and usually projecting from a wall to reinforce the wall.

10. _____ is heavy timber or a beam used to support vertical loads.

11. _____ is also called a vapor barrier, a material used to prevent the passage of moisture.

12. _____ is a rough frame used to form openings in poured concrete walls.

13. _____ is the depth to which the ground typically freezes in a particular area; footings must be placed below this depth.

14. _____ is a length of lumber or material applied over a joint to stiffen and strengthen it.

15. _____ is a strip of wood used to keep other pieces a desired distance apart.

Name: _____   Date: _____

Job Sheet 1: Estimating the Amount of Concrete for a Slab

• Upon completion of this job sheet, you will be able to estimate the amount of concrete for a slab.

Estimate the amount of concrete for the slab shown below. The depth of the slab is 6 inches.

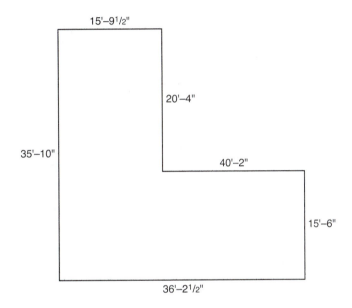

Instructor's Response

Name: _____   **Date:** _____

Job Sheet 2: Estimating the Amount of Concrete for a Footer

- Upon completion of this job sheet, you will be able to estimate the amount of concrete for a footer.

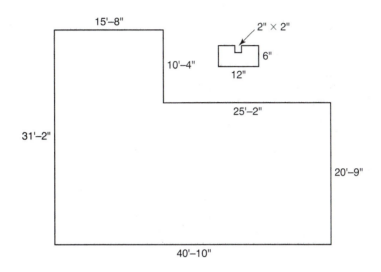

Instructor's Response

Chapter 7 Floor Framing

OBJECTIVES

Upon completion of this chapter, you will be able to:

Knowledge-Based

- ◎ Describe platform, balloon, and post-and-beam framing, and identify framing members of each.
- ◎ Describe several energy and material conservation framing methods.
- ◎ Describe methods to prevent destruction by wood pests.

Skill-Based

- ◎ Build and install girders, erect columns, and lay out sills.
- ◎ Lay out and install floor joists.
- ◎ Frame openings in floors.
- ◎ Lay out, cut, and install bridging.
- ◎ Apply subflooring.

Keywords

Anchor Bolt	Header	Sheathing
Balloon Frame	Joist	Shims
Band Joist	Linear Feet	Sill
Bridging	Masonry	Sill Sealer
Column	Plate	Subfloor
Dimension Lumber	Platform Frame	Tail Joists
Draftstops	Post	Termite Shields
Flush	Pressure-Treated	Trimmer
Girders	Ribbon	

Introduction

Once the foundation has been established and before the walls can be constructed, the floor must be framed and the subfloor material applied. After the floor has been framed and the subflooring installed, the walls can be constructed as well as the remaining components of the structure.

Chapter Review Questions and Exercises

COMPLETION

1. _____ is a type of frame in which studs are continuous from foundation sill plate to roof.

2. _____ is the member used to stiffen the ends of floor joists where they rest on the sill.

3. _____ are also called firestops; material used to reduce the size of framing cavities in order to slow the spread of fire; in a wood frame.

4. Dimension lumber blocking between studs is _____.

5. _____ is a term used to describe wood that is sold for framing and general construction.

6. _____ is the top or bottom horizontal member of a wall frame.

7. _____ is a large vertical member used to support a beam or girder.

8. _____ is a member placed at a right angle to joists, studs, and rafters to form and support openings in a wood frame.

9. _____ is a diagonal brace or solid wood block between floor joists used to distribute the load imposed on the floor.

10. _____ is a long metal fastener with a threaded end used to secure materials to concrete.

11. _____ is a treatment given to lumber that applies a wood preservative under pressure.

12. _____ is a term used to describe when surfaces or edges are aligned with each other.

13. _____ is a horizontal framing member used in a spaced pattern that provides support for the floor or ceiling system.

14. _____ is a heavy beam that supports the inner ends of floor joists.

15. _____ is a measurement of length.

16. _____ is any construction of stone, brick, tile, concrete, plaster, and similar materials.

17. _____ is a method of wood-frame construction in which walls are erected on a previously constructed floor deck or platform.

18. _____ is a vertical member used to support a beam or girder.

19. _____ are fastened to joist, rafters and studs on which the finish material is applied.

20. _____ is a narrow board let into studs of a balloon frame to support floor joists.

21. _____ is used as the first floor layer on top of joists.

22. _____ is material placed between the foundation and the sill to prevent air leakage.

23. _____ are shortened on center joists running from a header to a sill or girder.

24. _____ is a thin, wedge-shaped piece of material used behind pieces for the purpose of straightening them or for bringing their surfaces flush.

25. _____ is a metal flashing plate over the foundation to protect wood members from termites.

26. _____ is the first horizontal wood member resting on the foundation, supporting the framework of a building; also, the lowest horizontal member in a window or door frame.

27. _____ is a joist or stud placed at the sides of an opening running parallel to the main framing members.

Name: _____   Date: _____

Job Sheet 1: Determining the Joist Length of a Cantilevered Joist

- Upon completion of this job sheet, you will be able to determine the length of a cantilevered joist.

OVERALL DIMENSION	EXTEND OVER FOUNDATION

Overall Dimension	Extended Over Foundation
	4 feet
	6 feet
	3 feet
	5 feet–6 inches
	10 feet

Instructor's Response

Chapter 8   Wall and Ceiling Framing

OBJECTIVES

Upon completion of this chapter, you will be able to:

Knowledge-Based

- Identify and describe the function of each part of the wall frame.
- Determine the length of exterior wall studs.
- Describe four different types of walls used in residential framing.
- Determine the rough opening width and height for windows and doors.
- Describe several methods of framing corner and partition intersections.
- Describe the function of and install blocking and backing.
- Identify and describe the components of nonstructural steel wall framing.

Skill-Based

- Install a steel door buck.
- Estimate the materials needed for walls and ceiling framing.
- Apply wall sheathing.
- Lay out, cut, and install ceiling joists.
- Erect and temporarily brace a wall section plumb and straight.
- Assemble and construct a wall section.
- Lay out the wall plates for partition intersections, openings, and OC studs.

Keywords

Backing   Gypsum Board   Load Bearing
Blocking   Joist Hanger   Soffit
Gable End

Introduction

Once the foundation has been established and the floor framed and covered with plywood, it is time to start erecting the walls and then the roof. To frame a wall, you have to determine whether or not the wall is load bearing or not.

Wall Framing Components

The wall frame consists of a number of different parts. An exterior wall frame consists of the following components:

- Plates—top and bottom horizontal members of a wall frame

- Studs—vertical members of the wall frame

- Headers—run at right angles to studs

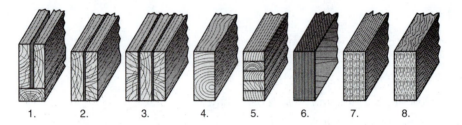

1. A BUILT-UP HEADER WITH A 2 X 4 OR 2 X 6 LAID FLAT ON THE BOTTOM.
2. A BUILT-UP HEADER WITH A 1/2" SPACER SANDWICHED IN BETWEEN.
3. A BUILT-UP HEADER FOR A 6" WALL.
4. A HEADER OF SOLID SAWN LUMBER.
5. GLULAM BEAMS ARE OFTEN USED FOR HEADERS.
6. A BUILT-UP HEADER OF LAMINATED VENEER LUMBER.
7. PARALLEL STRAND LUMBER MAKES EXCELLENT HEADERS.
8. LAMINATED STRAND LUMBER IS USED FOR LIGHT DUTY HEADERS.

FIGURE 8-1   Types of solid and built-up headers.

- Rough sills—form the bottom of a window opening at right angles to the studs

- Trimmers (jacks)—shortened studs that support the headers

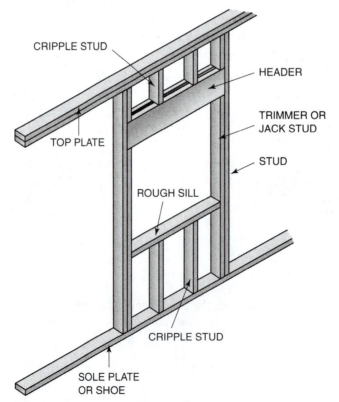

FIGURE 8-2   Typical framing for a window opening.

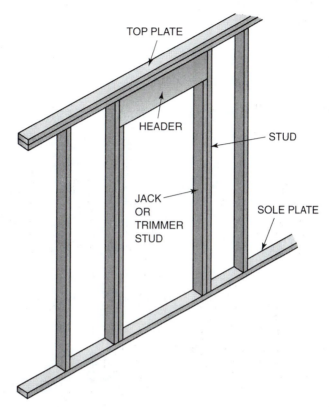

FIGURE 8-3 Typical framing for a door opening.

• Corner posts—same length as studs

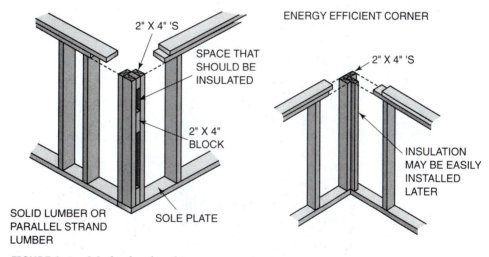

FIGURE 8-4 Methods of making corner posts.

- Partition intersections—framing needed when interior partitions meet an exterior partition

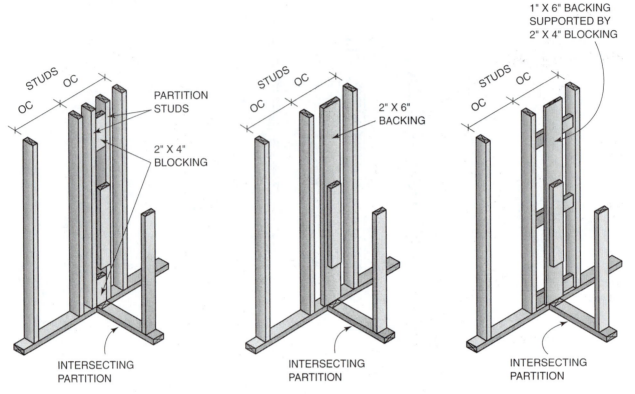

FIGURE 8-5 Partition intersections are constructed in several ways.

- Ribbons—horizontal members of the exterior wall frame in balloon construction

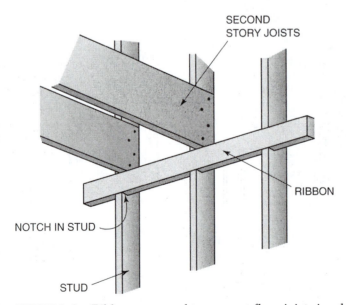

FIGURE 8-6 Ribbons are used to support floor joists in a balloon frame.

• Corner braces—used to brace walls

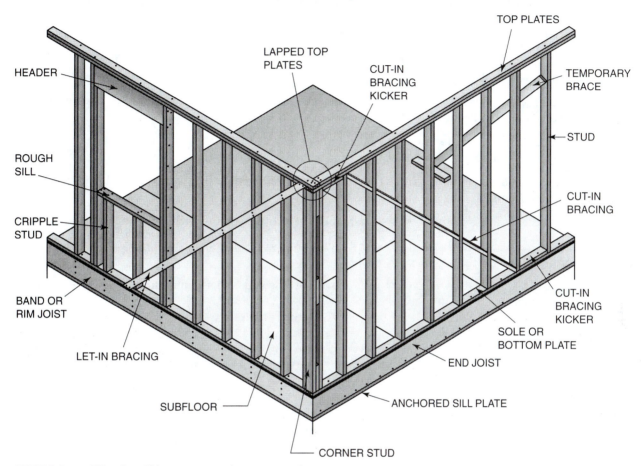

FIGURE 8-7 Wood wall bracing may be cut in or let in.

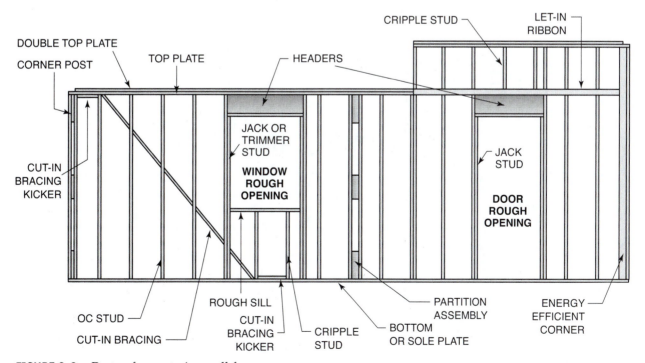

FIGURE 8-8 Parts of an exterior wall frame.

Chapter Review Questions and Exercises

COMPLETION

1. _____ is the triangular-shaped section on the end of a building formed by the common rafters and the top plate line.

2. _____ is a sheet product made by encasing gypsum in a heavy paper wrapping and is used to create the wall surface.

3. _____ are pieces of dimension lumber installed between joists and studs for the purposes of providing a nailing surface for intersecting framing members.

4. _____ is a term used to describe a structural member that carries weight from another part of the building.

5. _____ is the underside trim member of a cornice or any such overhanging assembly.

6. _____ are blocks of wood installed in walls or ceilings for the purpose of fastening or supporting trim or fixtures.

7. _____ is a metal stirrup used to support the ends of joists that do not rest on top of a support member.

Name: _____   Date: _____

Job Sheet 1: Types of Solid and Built-Up Headers

- Upon completion of this job sheet, you will be able to identify the various types of solid and built-up headers typically used in framing.

- Identify the types of headers pictured below.

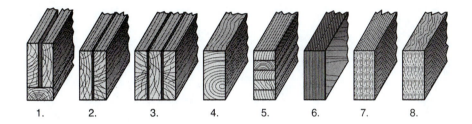

1.   2.   3.   4.   5.   6.   7.   8.

Instructor's Response:

Name: _____   **Date:** _____

Job Sheet 2: Window Framing

- Upon completion of this job sheet, you will be able to identify the various components of a window opening.

- Identify the components pictured below.

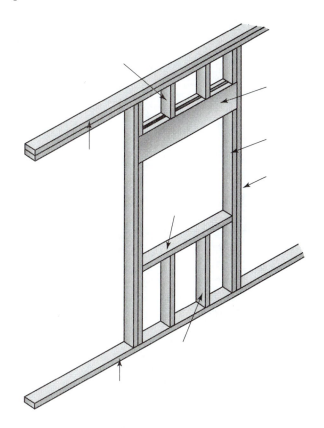

Instructor's Response

Name: _____   **Date:** _____

Job Sheet 3: Door Framing

- Upon completion of this job sheet, you will be able to identify the various components of a door opening.

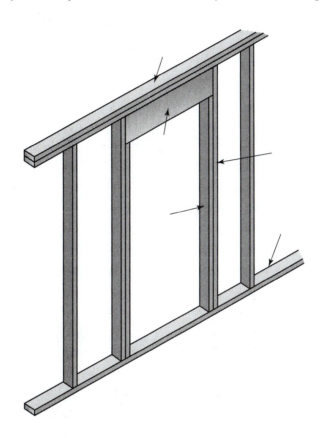

Instructor's Response

Scaffolding, Ladders, and Sawhorses

Upon completion of this chapter, you will be able to:

Knowledge-Based

- Identify and describe the safety concerns for scaffolds.
- Describe the recommended capacities of various parts of a scaffold.
- Identify and describe the components of a fall protection system.
- Describe the safety concerns for mobile metal tubular scaffolds.
- Describe the safe use of ladders, ladder jacks, stepladders, and sawhorses.

Skill-Based

- Build safe staging areas by using roof brackets.
- Safely set up, use, and dismantle pump jack scaffolding.
- Construct a scaffold work platform.
- Erect and dismantle metal scaffolding in accordance with recommended safety procedures.
- Follow a recommended procedure to inspect a scaffold for safety.

Keywords

Cleat	Crib	Users
Competent Person	Erectors	

Introduction

Ladders can be divided into two main types, straight and step. Straight ladders are constructed by placing rungs between two parallel rails. They generally contain safety feet on one end that help prevent the ladder from slipping.

Step ladders are self-supporting, constructed of two sections hinged at the top. The front section has two tails and steps; the rear portion has two rails and braces.

Safe Ladder Placement

- Ladders, including step ladders, shall be placed so that each side rail (or stile) is on a level and firm footing and so that the ladder is rigid, stable, and secure.

- The stiles shall not be supported by boxes, loose bricks, or other loose packing.

- No ladder shall be placed in front of a door opening toward the ladder unless the door is fastened open, locked, or guarded.

- Straight ladders should be placed against the side of a building or other structure at an angle of approximately 76°.

- Where a ladder passes through an opening in the floor of a landing place, the opening shall be as small as possible.

- A ladder placed such that its top end rests against a window frame shall have a board fixed to its top end. The size and position of this board shall ensure that the load to be carried by the ladder is evenly distributed over the window frame.

Safely Securing Ladders

- Ladders shall be securely fixed at the top and foot so that they cannot move either from their top or from their bottom points of rest. If it is not possible to secure a ladder at both the top and bottom, then it shall be securely fixed at the base. If this is not possible, then a person should stand at the base of the ladder and secure it manually against slipping.

- Ladders set up in public thoroughfares or other places (where there is potential for accidental collision with them) must be provided with effective means to prevent the displacement of the ladders due to collisions (for example, use of barricades).

Safe Use of Ladders

- Only one person at a time may use or work from a single ladder.

- Always face the ladder when ascending or descending it.

- Carry tools in a tool belt, pouch, or holster so you can keep hold of the ladder.

- Wear fully enclosed slip resistant footwear when using ladder.

- Do not climb higher than the third rung from the top of the ladder.

- Ladders made by fastening cleats across a single rail or stile shall not be used.

- When there is significant traffic on ladders used for building work, separate ladders for ascent and descent shall be provided, designated, and used.

- Make sure the weight your ladder is supporting does not exceed its maximum load rating (user plus materials). There should only be one person on the ladder at one time.

- Use a ladder that is the proper length for the job. Proper length is a minimum of 3 feet extending over the roofline or working surface. The three top rungs of a straight, single, or extension ladder should not be stood on.

- Straight, single, or extension ladders should be set up at about a 76° angle.

- Metal ladders will conduct electricity. Use a wooden or fiberglass ladder in the vicinity of power lines or electrical equipment. Do not let a ladder made from any material contact live electric wires.

- Be sure all locks on extension ladders are properly engaged.

- The ground under the ladder should be level and firm. Large flat wooden boards braced under the ladder can level a ladder on uneven ground or soft ground. A good practice is to have a helper hold the bottom of the ladder.

- The two top rungs are not for standing on a step ladder.

- Follow the instruction labels on ladders.

Basic Fall Protection Safety Procedures

Maintaining written fall protection procedures not only protects workers from falls, but also protects management from charges of incompetence. Having individual workers or supervisors decide when fall protection is required and what kinds of fall protection equipment to use is an acceptable practice only where workers are routinely exposed to simple hazards—homebuilders on a roof, for example. However, when workers are involved with lots of non-routine jobs, safety is enhanced if management puts the fall protection and rescue procedures that employees are required to use in writing.

The written plan must describe how workers will be protected when working 10 feet or more above the ground, other work surfaces, or water.

The plan should:

1. Identify all fall hazards in the work area.

2. Describe the method of fall arrest or fall restraint to be provided.

3. Outline the correct procedures for the assembly, maintenance, inspection, and disassembly of the fall protection system to be used.

4. Explain the method of providing overhead protection for workers who may be in or pass through the area below the work site.

5. Communicate the method for prompt, safe removal of injured workers.

Before a fall protection plan can be developed, two important definitions must be understood:

1. Fall arrest system: This equipment protects someone from falling more than 6 feet or from striking a lower object in the event of a fall, whichever distance is less. This equipment includes approved full-body harnesses and lanyards properly secured to anchorage points or to lifelines, safety nets, or catch platforms.

2. Fall restraint system: This apparatus keeps a person from reaching a fall point; for example, it allows someone to work up to the edge of a roof but not fall. This equipment includes standard guardrails, a warning line system, a warning line and monitor system, and approved safety belts (or harnesses) and lanyards attached to secure anchorage points.

Developing a Fall Protection Work Plan

To develop a fall protection work plan, you must identify the responsibilities of your company and the work areas to which the plan applies. This information should be listed as the first item in your plan. After you identify your company responsibilities and work areas in the plan:

1. Identify all fall hazards in the work area. To determine fall hazards, you must review all jobs and tasks to be done. After all fall hazards have been identified, list those requiring employees to work 10 feet or more above the ground, other work surface, or water.

2. Determine the method of fall arrest or fall restraint to be provided for each job and task to be done that is 10 feet or more above the ground, another work surface, or water.

3. Describe the procedures for assembly, maintenance, inspection, and disassembly of the fall protection system to be used.

4. Describe the correct procedures for handling, storage, and security of tools and materials.

5. Describe the method of providing overhead protection for workers who may be in or pass through the area below the work site.

6. Describe the method for prompt, safe removal of injured workers.

7. Include where a copy of this plan will be posted.

8. Train and instruct all personnel in all of the above items.

9. Keep a record of employee training and maintain it on the job.

Chapter Review Questions and Exercises

MULTIPLE CHOICE

1. The base of a ladder should be 1 foot away from the vertical surface against which it is placed for every _____ feet in height.

   a. 2   c. 4

   b. 3   d. 5

2. Employees must be able to safely access any level of scaffold that is _____ feet above or below an access point.

   a. 2   c. 4

   b. 3   d. 5

3. Fall protection must be provided for workers on scaffolds when they are _____ feet or more above a lower level.

   a. 6   c. 12

   b. 10   d. 15

4. When using ladders near electrical equipment, ladders made of _____ should be used.

   a. wood   c. fiberglass

   b. aluminum   d. either A or C

5. The most common type of scaffolding used in the construction industry is the _____.

   a. pump jack   c. moveable

   b. fabricated frame   d. adjustable

SHORT ANSWER

6. Explain what a yellow tag on a scaffold means.

   _____

   _____

   _____

   _____

7. Explain what a red tag means when it is attached to a scaffolding.

   _____

   _____

   _____

   _____

COMPLETION

8. _____ is a designated person on a job site who is capable of identifying hazardous or dangerous situations and has the authority to take prompt corrective measures to eliminate them.

9. _____ is a worker whose responsibilities include safe assembly of scaffolding.

10. _____ is a small strip of wood applied to support a shelf or similar piece.

11. _____ are people who work on scaffolding.

12. _____ is heavy wood blocks and framing used as a foundation for scaffolding.

Name: _____   **Date:** _____

Job Sheet 1: Fall Protection Plan

- Upon completion of this job sheet, you will be able to create a fall protection plan.
- You are the construction manager of a high-rise complex. As the manager, you are required to develop a fall protection plan for a group of carpenters working on the tenth floor. The facility is a 35-story building (385 feet) in which the facade is mostly glass. The framing crew consists of three people replacing a portion of the outside wall.

Plan

Instructor's Response

Chapter 10 Roof Framing

OBJECTIVES

Upon completion of this chapter, you will be able to:

Knowledge-Based

- ⊙ Describe several roof types.
- ⊙ Define the various roof framing terms.
- ⊙ Identify the members of gable, gambrel, hip, intersecting, and shed roofs.
- ⊙ Describe and perform the safe and proper procedure to erect a trussed roof.

Skill-Based

- ⊙ Lay out a common rafter and erect a gable roof.
- ⊙ Lay out and install gable end studs.
- ⊙ Lay out a hip rafter and hip jack rafters.
- ⊙ Lay out valley rafters.
- ⊙ Apply roof sheathing.
- ⊙ Estimate the quantities of materials used in a roof frame.

Keywords

Cheek Cut	Hip-Valley Cripple Jack Rafter	Shed Roof
Dormer	Intersecting Roof	Tail Cut
Fascia	Lateral	Valley
Gable Roof	Lookout	Valley Cripple Jack Rafter
Gambrel Roof	Mansard Roof	Valley Jack Rafter
Hip Jack	Rake	Valley Rafter
Hip Rafter		

Introduction

When a roof is being planned, the angle of the roof is an important consideration and special care must be taken to ensure that the roof has adequate slope to allow for proper drainage. The angle of the roof's slope is described by using the ratio known as *pitch*. Pitch is the ratio of the vertical rise to the horizontal run—Vertical Rise/Horizontal Run. For example, a roof with a pitch of 6:12 increases vertically six inches for every horizontal foot. To calculate pitch, the following formula is used.

Rise/Run = Pitch/12

Therefore, Pitch = Rise × 12/Run

And is expressed as Pitch:12

For example, if the horizontal run of a roof is 12 feet and the vertical rise is 4 inches, then the pitch of the roof can be calculated by using the formula:

Pitch = Rise × 12/Run
Pitch = 4 inches (12)/12
Pitch = 48/12
Pitch = 4:12

Chapter Review Questions and Exercises

COMPLETION

1. _____ is a type of roof that has two slopes of different pitches on each side of its center.

2. _____ is a compound miter cut on the end of certain roof rafters.

3. _____ is a rafter running between a hip rafter and the wall plate.

4. _____ is a structure that projects out from a sloping roof to form another roofed area to provide a surface for the installation of windows.

5. _____ is a common type of roof that pitches in two directions.

6. _____ is a vertical member of the cornice finish installed on the bottom end of rafters.

7. _____ extends diagonally from the corner of the plate to the ridge at the intersection of two surfaces of a roof.

8. _____ is a rafter running between two valley rafters.

9. _____ is a rafter running between a valley rafter and the ridge.

10. _____ is a direction to the side at about 90°.

11. _____ is horizontal framing pieces in a cornice, installed to provide fastening for the soffit.

12. _____ is the roof of irregular-shaped buildings; valleys are formed at the intersection of the roofs.

13. _____ is the rafter placed at the intersection of two roof slopes in interior corners.

14. _____ is the sloping portion of the gable ends of a building.

15. _____ is the intersection of two roof slopes at interior corners.

16. _____ is a cut on the extreme lower end of a rafter.

17. _____ is a type of roof that slopes in one direction only.

18. _____ is a type of roof that has two different pitches on all sides of the building, with the lower slopes steeper than the upper slopes.

19. _____ is a short rafter running parallel to common rafters, cut between hip and valley rafters.

Name: _____   Date: _____

Job Sheet 1: Truss Component Identification

- Upon completion of this job sheet, you will be able to identify the various trusses used in a residential construction.
- For each of the trusses listed create a quick sketch.

Scissors Truss

Modified Fink

Camel Back Pratt

Saw-Tooth Truss

Instructor's Response

Name: _____   **Date:** _____

Job Sheet 2: Truss Component Identification

- Upon completion of this job sheet, you will be able to identify the various components of a roof truss.
- Identify the various components of the truss pictured below.

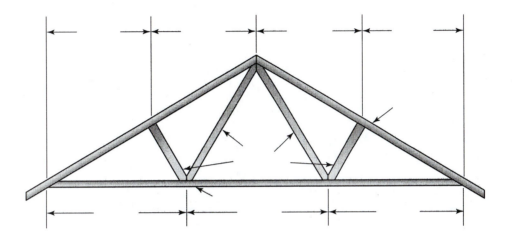

Instructor's Response

Chapter 11 Windows and Doors

Upon completion of this chapter, you will be able to:

Knowledge-Based

- Describe the most popular styles of windows and name their parts.
- Name the parts of and set a prehung door frame.
- Describe the standard designs and sizes of doors and name their parts.

Skill-Based

- Fit and hang a door to a pre-existing opening.
- Install locksets in doors.
- Install bypass, bifold, and pocket doors.
- Select and specify desired sizes and styles of windows from manufacturers' catalogs.
- Install various types of windows in an approved manner.

Keywords

Astragal	Glazing	Rail
Back Miter	Hopper Window	Sash
Bay Window	House Wrap	Stile
Casing	Insulated Glass	Strike Plate
Deadbolt	Low Emissivity Glass	Weather Stripping
Double-Acting	Molding	Wind
Escutcheon	Mullion	
Extension Jambs	Muntin	

Introduction

Doors are used to provide security, privacy, and fire protection to access openings in interior and exterior walls. The choice of a door depends on the traffic passing through the opening and the door's appearance. In residential work, interior doors provide privacy and some degree of security, and exterior doors provide protection from the weather, security, and, often, an important part of the exterior design. In commercial buildings, doors must permit ingress and egress as required by building codes. They are subject to considerable use, so they must be strong and have hardware that can withstand constant use. Other doors provide access for large equipment, such as trucks and aircraft. They are usually mechanically opened and closed because they are large and heavy. Doors are also used to divide large interior spaces into smaller rooms.

Various windows are manufactured primarily for use in residential buildings. Windows made from solid wood, wood clad with plastic or aluminum, solid plastic, steel, stainless steel, aluminum, bronze, and composite materials are available.

A major consideration in the selection of windows is energy savings. Various types of energy-efficient glazing are available, such as energy-efficient glass, double glazing, and glass with louver blinds set between the panes of glass. Some double glazed units have a gas inserted in the space between the panes.

Installing an Interior Door

The following steps outline the procedure for installing an interior door:

1. Remove the protective packing from the unit.

2. Center the unit in the opening so that the door will swing in the desired direction.

3. Level the head jambs.

4. Punch the hinge side of the door unit.

5. Open the door and move to the other side. (Check that the door is nearly centered.)

6. Check the operation of the door, making any necessary adjustments.

7. Finish nailing the casing and install casing on the other side of the door.

Re-Key Locks

The following steps outline the procedure for re-keying a lock:

1. Take the locking part of the lock out by unscrewing it from the inside. Leave the rest of the lock in place.

2. Take the knob with the locking mechanism with its proper key to the hardware store. The hardware store employee will replace the pins so they'll fit another key.

3. Put the door lock back together.

Installing a Window

The following steps outline the procedure for installing a window:

1. Install housewrap by beginning at the building corner holding the roll vertical on the wall.

2. Make cuts in the housewrap from corner to corner of the rough opening.

3. Place window in the opening after removing all shipping protection from the window unit.

4. Center the unit in the opening on the rough sill with the exterior window casing against the wrapping wall sheathing.

5. Remove the window unit from the opening and caulk the back side of the casting or nailing flange.

6. Flash the head casing by cutting the flashing to length with tin snips.

Chapter Review Questions and Exercises

COMPLETION

1. _____ is a type of window in which the sash is hinged at the bottom and swings inward.

2. _____ is a semicircular molding often used to cover a joint between two doors.

3. _____ is a coating on double glazed windows designed to raise the insulating value by reflecting heat back into the room.

4. _____ is a decorative strip of wood used for finishing purposes.

5. _____ is a slender strip of wood between lights of glass in windows or doors.

6. _____ is the horizontal member of a frame.

7. _____ is that part of a window into which the glass is set.

8. _____ is an angle cut starting from the end and going back on the face of the stock.

9. _____ is a protective plate covering the knob or key hole in doors.

10. _____ is a window, usually three-sided, that projects out from the wall line.

11. _____ is a molding used to trim around doors, windows, and other openings.

12. _____ are doors that swing in both directions, or the hinges used on these doors.

13. _____ is the act of installing glass in a frame.

14. _____ is a vertical division between window units or panels in a door.

15. _____ is a pane of glass or an opening for a pane of glass.

16. _____ is a type of building paper with which the entire exterior sidewalls of a building are covered.

17. _____ is a strip of wood added to window jambs to bring the jamb edge flush with the wall surface in preparation for casing.

18. _____ are multiple panes of glass fused together with an air space between them.

19. _____ is the outside vertical members of a frame, such as in a paneled door.

20. _____ is a narrow strip of material applied to windows and doors to prevent the infiltration of air and moisture.

21. _____ is a defect in lumber caused by a twist in the stock from one end to the other; also, a twist in anything that should be flat.

22. How are items of finish carpentry and millwork generally measured?

_____

_____

_____

23. Describe how doors and doorframes are measured.

_____

_____

_____

Name: _____   Date: _____

Job Sheet 1: Hanging an Interior Door

- Upon completion of this job sheet, you will be able to identify the steps for hanging an interior door.
- In the space provided below, list the steps for hanging an interior door.

Procedure

Instructor's Response

Name: _____   **Date:** _____

Job Sheet 2: Installing Windows

- Upon completion of this job sheet, you will be able to identify the steps for installing windows.
- In the space provided below, list the steps for installing windows.

Procedure

Instructor's Response

Chapter 12 Roofing

OBJECTIVES

Upon completion of this chapter, you will be able to:

Knowledge-Based

- ⊗ Define roofing terms.
- ⊗ Describe and apply roofing felt underlayment, organic or fiberglass asphalt shingles, and roll roofing.
- ⊗ Describe and apply flashing to valleys, sidewalls, chimneys, and other roof obstructions.

Skill-Based

- ⊗ Estimate needed roofing materials.

Keywords

Apron	Electrolysis	Saddle
Asphalt Felt	Exposure	Selvage
Closed Valley	Flashing	Square
Cricket	Mortar	
Drip Edge	Open Valley	

Introduction

The type of material used for weather protection must be specified on the roof plan. Common roofing materials include single-ply and built-up roofs, shingles, metal slate, and tile roofs. Many roof materials are also described by their weight per square. A *square* of roofing is equal to 100 square feet (9.3 m^2).

Single-ply roofs can be applied as a thin liquid or sheet made from *ethylene propylene diene monomer (EPDM)*, which is an elastomeric or synthetic rubber material. Polyvinyl chloride (PVC), chlorosulfonated polyethylene (CSPE), and polymer-modified bitumens are also used. These materials are designed to be applied to roof decks ranging with a minimum slope 1/4:12. In liquid form, these materials can be applied directly to the roof decking with a roller or sprayer to conform to irregular-shaped roofs.

Single-sheet roofs are rolled out and bonded together to form one large sheet. The sheet can be bonded to the roof deck by mechanical fasteners. Some applications are not attached to the roof deck and are held in place by gravel material that is placed over the roofing material to provide ballast. A typical specification for a single-ply roof would specify the material, the application method, and the aggregate size. Many firms also specify a reference for the material to be installed according to the manufacturer's specification.

Repairing Roofing

The following steps outline the process for installing asphalt shingles:

1. Prepare the roof deck by cleaning sawdust and debris that will cause a slipping hazard.
2. Begin underlayment over the deck at a lower corner.
3. Nail or staple through each lap and through the center of each layer about 16 inches apart.
4. Install metal drip edge along the perimeter on top of the underlayment.
5. Prepare the starter course by cutting off the exposed laps lengthwise through the shingle.
6. Determine the starting line, either the rake edge or vertical center snapped lines.
7. Starting shingles layout at the rake edge involves placing the first course, with a whole tab at the rake edge.
8. If the cutouts are to break on the third, cut the starting strip for the second course by removing 4 inches.
9. Fasten each shingle from the end nearest the shingle just laid.
10. Install vented ridge cap per manufacturer's instructions.
11. Apply the cap across the ridge until 3 or 4 feet from the end.

The following steps outline the process for installing roll roofing:

1. Apply 9-inch-wide strips of roofing along the eaves and rakes overhanging the drip edge about ⅜ inch.
2. Apply the first course of roofing with its edge and ends flush with the strips.
3. Apply cement only to the edge strips covered by the first course.
4. Apply succeeding course in like manner.
5. After all courses are in place, lift the lower edge of each course.
6. To cover the hips and ridge, cut strips of 12 × 36-inch roofing.
7. Snap a chalk line on both sides of the hip or ridge down about 5½ inches from the center.
8. Press it firmly into place.
9. Spread cement on the end of each strip that is lapped before the next one is applied.

The following steps outline the procedure for installing double coverage roll roofing:

1. Cut the 19-inch strip of selvage non-mineral surface side from enough double coverage roll roofing to cover the length of the roof.
2. Apply the first course by using a full-width strip of roofing.
3. Apply succeeding courses in the same manner.
4. Lift and roll back the surface portion of each course.
5. Apply the remaining surfaced portion left from the first course as the last course.
6. It is important to follow a specific application's instructions because of differences in the manufacturing of roll roofing.

The following steps outline the procedure for installing woven valley:

1. Install underlayment and starter strip to both roofs.
2. Apply the first course on one roof, the left one, into and past the center of the valley.

3. Apply the first course of the other (right) roof in a similar manner, into and past the valley.

4. Succeeding courses are applied by repeating this alternating pattern, first from one roof and then on the other.

The following steps outline the procedure for installing closed cut valley:

1. Begin by shingling the first roof completely, lettering the end shingle of every course overlap the valley by at least 12 inches.

2. Snap a chalk line along the center of the valley on top of the shingles of the first roof.

3. Apply the shingles of the second roof, cutting the end shingles of each course to the chalk line.

The following steps outline the procedure for installing step flashing:

1. Snap a chalk line in the center of the valley on the valley underlayment.

2. Apply the shingle starter course on both roofs.

3. Fit and form the first piece of flashing to the valley on top of the starter strips.

4. Apply the first regular course of shingles to both roofs on each side of the valley, trimming the ends to the chalk line.

Chapter Review Questions and Exercises

COMPLETION

1. _____ is the unexposed part of roll roofing covered by the course above.

2. _____ is a mixture of Portland cement, lime, sand, and water and is used to bond masonry units together.

3. _____ is a building paper saturated with asphalt for waterproofing.

4. _____ a roof valley in which the roof covering is kept back from the centerline of the valley.

5. _____ is the amount that courses of siding or roofing are exposed to the weather.

6. _____ is the amount of roof covering that will cover 100 square feet of roof area.

7. _____ is the flashing piece located on the lower side of a roof penetration such as a chimney or dormer.

8. _____ is the material used at intersections such as roof valleys and dormers and above windows and doors to prevent the entrance of water.

9. _____ is a roof valley in which the roof covering meets in the center of the valley, completely covering the valley.

10. _____ is a metal edging strip placed on roof edges to provide a support for the overhang of the roofing material.

11. _____ is accelerated oxidation of one metal because of contact with another metal in the presence of water.

Name: _____   **Date:** _____

Job Sheet 1: Estimating Shingles

- Upon completion of this job sheet, you will be able to estimate the number of shingles needed to roof a house.

- Estimate the number of shingles needed to roof a house if the fascia boards measure 55 feet and the length of the rafter is 25 feet.

Instructor's Response

Name: _____   **Date:** _____

Job Sheet 2: Installing Asphalt Shingles

- Upon completion of this job sheet, you will be able to describe the method used to install asphalt shingles.
- In the space provided below, list the steps for installing asphalt shingles.

Instructor's Response

Chapter 13 Siding and Decks

OBJECTIVES

Upon completion of this chapter, you will be able to:

Knowledge-Based

- Describe the shapes, sizes, and materials used as siding products.
- Describe various types of cornices and name their parts.
- Describe the construction of and kinds of materials used in decks.

Skill-Based

- Lay out and install footings, supporting posts, girders, and joists for a deck.
- Apply decking in the recommended manner and install flashing for an exposed deck against a wall.
- Install corner boards and prepare sidewall for siding.
- Apply horizontal and vertical siding.
- Apply plywood and lapped siding.
- Apply wood shingles and shakes to sidewalls.
- Apply vinyl and aluminum siding.
- Install gutters and downspouts.

Keywords

Baluster	Drip	Soffit
Battens	Eaves	Story Pole
Blind Nail	Frieze	Striated
Corner Boards	Gutter	Water Table
Cornice	Plancier	
Downspout	Rake	

Introduction

Additional living space can be created by utilizing a portion of the exterior of the home by constructing decks. Like kitchen cabinets (which will be discussed in Chapter 17) decks are a reflection of the personality of the homeowner. It showcases their style and taste. Decks can be designed in a variety of styles, materials, and shapes. Another reflection of the homeowner's personality is the exterior covering of the home itself (for example, wood, vinyl, or aluminum siding).

Estimating the Amount of Aluminum and Vinyl Siding

Aluminum and vinyl siding panels are typically sold by the square. To determine the amount of siding needed for a project, first determine the amount of wall area to be covered. Second, add 10% of the wall area calculated for waste. Third, divide the amount calculated in the previous step by 100. For example, to determine the amount of siding needed for the medical offices shown below, having an exterior wall height for the structure is 9 feet 6 inches.

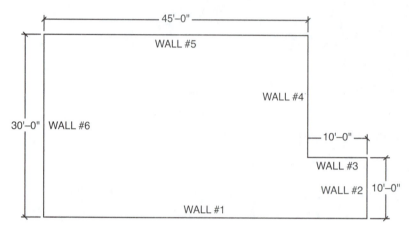

FIGURE 13-1 Estimating siding.

1. Determine the wall area to be covered.

   Wall #1

   Length = length of wall #5 + length wall #3
   Length = 45'0" + 10'0"
   Length = 55'0"

   Because the height of the wall is 9'6" or 9.5', the area is calculated as

   Area1 = 55' · 9.5'
   Area1 = 522.5 sq. ft.

   Wall #2

   Area2 = 10'0" · 9.5'
   Area2 = 95 sq. ft.

   Wall #3

   Area3 = 10'0" · 9.5'
   Area3 = 95 sq. ft.

   Wall #4

   Area4 = 20'0" · 9.5'
   Area4 = 190 sq. ft.

   Wall #5

   Area5 = 45'0" · 9.5'
   Area5 = 427.5 sq. ft.

Wall #6

Area6 = 30′0″ · 9.5′

Area6 = 285 sq. ft.

Total Square Footage

Area Total = Area1 + Area2 + Area3 + Area4 + Area5 + Area6

Area Total = 522.5 sq. ft. + 95 sq. ft. + 95 sq. ft. + 190 sq. ft. + 427.5 sq. ft. + 285 sq. ft.

Area Total = 1615 sq. ft.

2. Determine the amount of waste.

Allowing for waste

Waste Allowance = Area Total · 0.10

Waste Allowance = 161.5 sq. ft.

3. Determine the number of squares.

Total Wall Area with Waste = 1615 sq. ft. + 161.5 sq. ft.

Total Wall Area with Waste = 1776.5 sq. ft.

Number of Square = Total Wall Area with Waste/100

Number of Square = 1778.5/100

Number of Square = 17.765 or 18

Repairing Siding

The following steps outline the procedure for installing horizontal siding.

1. First determine the siding exposure so that it is about equal both above and below the window sill.

2. Install a starter strip of the same thickness and width of the siding at the headlap fastened along the bottom edge of the sheathing.

3. From this first chalk line, lay out the desired exposures on each corner board and each side of all openings.

4. Install the siding as per the manufacturer's recommendations, staggering the butt joints in adjacent courses as far apart as possible.

5. When applying a course of siding, start from one end and work toward the other end.

6. Siding is fastened to each bearing stud or about every 16 inches.

The following steps outline the procedure for installing vertical tongue-and-groove roofing.

1. Slightly back-bevel the ripped edge.

2. Fasten a temporary piece on the other end of the wall projecting below the sheathing by the same amount.

3. Apply succeeding pieces by toe-nailing into the tongue edge of each piece.

4. To cut the piece to fit around an opening, first fit and tack a siding strip in place where the last full strip will be located.

5. Next, use a scrap block of the siding material, about 6 inches long, with the tongue removed.

6. Continue the siding by applying the short lengths across the top and bottom of the opening as needed.

7. Fit the next full-length siding piece to complete the siding around the opening.

8. Remove the piece and the scrap blocks from the wall.

The following steps outline the procedure for installing panel siding.

1. Install the first piece with the vertical edge plumb.
2. Apply the remaining sheets in the first course in like manner.

The following steps outline the procedure for installing wood shingles and shakes.

1. Fasten a shingle on both ends of the wall with its butt about 1 inch below the top of the foundation.
2. Fill in the remaining shingles to complete the undercourse.
3. Apply another course on top of the first course.
4. To apply the second course, snap a chalk line across the wall at the shingle butt line.

The following steps outline the procedure for installing horizontal vinyl siding.

1. Snap a level line to the height of the starter strip all around the bottom of the building.
2. Cut the corner posts so they extend ¼ inch below the starting strip.
3. Cut each j-channel piece to extend, on both ends, beyond the casing and sills a distance equal to the width of the channel face.
4. On both ends of the top and bottom channels, make ¾-inch cuts at the bends leaving the tab attached.
5. Snap the bottom of the first panel into the starter strip.
6. Install successive courses by interlocking them with the course below and staggering the joints between courses.
7. To fit around a window, mark the width of the cutout, allowing ¼-inch clearance on each side.
8. Panels are cut and fit over windows in the same manner as under them.
9. Install the last course of the siding panel under the soffit in a manner similar to fitting under a window.

The following steps outline the procedure for installing vertical vinyl siding.

1. Measure and lay out the width of the wall section for siding pieces.
2. Cut the edge of the first panel nearest the corner.
3. Install the remaining full strips, making sure there is ¼ inch gap at the bottom.

Chapter Review Questions and Exercises

COMPLETION

1. _____ is the part of an exterior finish that projects below another to cause water to drop off instead of running back against and down the wall.

2. _____ is a vertical member used to carry water from the gutter downward to the ground; also called leader.

3. _____ is the horizontal, underside trim member of a cornice or other overhanging assembly.

4. _____ is a finish material with random and finely spaced grooves running with the grain.

5. _____ is a method of fastening that conceals the fastener.

6. _____ is a narrow strip of wood used to lay out the insulation heights of material such as siding or vertical members of a wall frame.

7. _____ is a thin, narrow strip typically used to cover joints in vertical boards.

8. _____ are finish trim members used at the intersection of exterior walls.

9. _____ is the vertical member of a stair rail usually decorative and spaced closely together.

10. _____ is the finish member on the underside of a box cornice, also called a soffit.

11. _____ is a general term used to describe the part of the exterior finish where the walls meet the roof.

12. _____ is the lower part of the roof that extends beyond the sidewalls.

13. _____ is that part of the exterior trim applied to cover the joint between the overhanging cornice and the siding.

14. _____ is a trough attached to an eave used to carry off water.

15. _____ is the portion of the roof that overhangs the gable end.

16. _____ is exterior trim members applied at the intersection of the siding and the foundation that projects outward to direct water away from the building.

Name: _____   Date: _____

Job Sheet 1: Installing Vinyl Siding

- Upon completion of this job sheet, you will be able to describe the method used to install vinyl siding.
- In the space provided below, list the steps for installing vinyl siding.

Instructor's Response

Name: _____   Date: _____

Job Sheet 2: Installing Wood Shingles and Shakes

- Upon completion of this job sheet, you will be able to describe the method used to install wood shingles and shakes.
- In the space provided below, list the steps for installing wood shingles and shakes.

Instructor's Response

Chapter 14 Insulation and Wall Finish

OBJECTIVES

Upon completion of this chapter, you will be able to:

Knowledge-Based

- ⊗ Describe how insulation works and define insulating terms and requirements.
- ⊗ Describe the commonly used insulating materials and state where insulation is placed.
- ⊗ State the purpose of and install vapor retarders.
- ⊗ Explain the need for ventilating a structure and describe types of ventilators.
- ⊗ Describe various kinds, sizes, and uses of gypsum panels.
- ⊗ Describe the kinds and sizes of nails, screws, and adhesives used to attach gypsum panels.
- ⊗ Describe and apply several kinds of sheet wall paneling.
- ⊗ Describe and apply various patterns of solid lumber wall paneling.

Skill-Based

- ⊗ Estimate quantities of drywall, drywall accessories, and sheet and board wall paneling.
- ⊗ Properly install various kinds of insulation.
- ⊗ Apply gypsum board to interior walls and ceilings.
- ⊗ Conceal gypsum board fasteners and corner beads.
- ⊗ Reinforce and conceal joints with tape and compound.

Keywords

Air Infiltration	Gypsum Board	Vapor Retarder
Condensation	Hardboard	Wainscoting
Dew Point	Insulation	Weather Stripping
Eased Edge	R-Value	
Face	Storm Sash	

Introduction

Once the exterior walls have been covered and the structure is dry, then the interior walls can be covered. In many cases, this involves covering the interior walls with drywall. Today, drywall is the material of choice because it can be easily cut and installed with very little training. In addition, it can be easily transformed into almost any shape. Finally, it provides the homeowner with a wall surface that can be easily painted, wallpapered, and so on.

Estimating

The following steps outline the procedure for estimating the amount of drywall material needed.

1. Determine the area of the walls and ceiling to be covered.

   Ceiling—multiply the length of the room by its width

   Walls—multiply the perimeter of the room by the height

2. Subtract only the large wall openings, such as double doors.

3. Combine all areas to find the total number of square feet of drywall.

4. Add about 5% of the total for waste.

5. Divide the total area to be covered by the area of one panel to get the number of panels.

Installing Drywall

The following steps outline the procedure for cutting and fitting gypsum board.

1. Take measurements accurately to within ¼ inch for the ceiling and ⅛ inch for the walls.

2. Bend the board back against the cut.

3. Lifting the panel off the floor, snap the cut piece back quickly to a straight position.

4. To make cuts parallel to the long edges, the board is often gauged with a tape and scored with a utility knife.

5. Ragged edges can be smoothed with a drywall rasp, a coarse sanding block, or a knife.

Installing Sheet Paneling

The following steps outline the procedure for installing sheet paneling.

1. Mark the location of each stud in the wall on the floor and ceiling.

2. If the wall is to be wainscoted, snap a horizontal line across the wall to indicate its height.

3. Apply narrow strips of paint on the wall from floor to ceiling over the stud where a seam in the paneling will occur.

4. Cut the first sheet to a length of about ¼ inch less than the wall height.

5. Notice the joint at the corner and the distance the sheet edge overlaps the stud.

6. Remove the sheet from the wall and cut close to the scribed line.

7. If a tight fit between the panel and ceiling is desired, set the dividers and scribe a small amount at the ceiling line.

Installing Wall and Ceiling Insulation

The following steps outline the procedure for installing flexible insulation.

1. Install positive ventilation chutes between the rafters where they meet the wall plate.

2. Install the air-insulation dam between rafters in line or on the exterior sheathing.

3. To cut the material, place a scrap piece of plywood on the floor to protect the floor while cutting.

4. Place the batts or blankets between the studs.

5. Fill any spaces around windows and doors with spray-can foam.

6. Install ceiling insulation by stapling it to the ceiling joist or by friction-fitting it between them.

7. Flexible insulation installed between floor joists over crawl spaces may be held in place by wire mesh or pieces of heavy gauge wire wedged between the joists.

Chapter Review Questions and Exercises

COMPLETION

1. _____ is a panel used as a finished surface material made from a mineral mined from the earth.

2. _____ is when water, in a vapor form, changes to a liquid due to cooling of the air; the resulting droplets of water that accumulate on the cool surface.

3. _____ is a material used to restrict the passage of heat or sound.

4. _____ is the temperature at which moisture begins to condense out of the air.

5. _____ is a building product made by compressing wood fibers into sheet form.

6. _____ is a wall finish applied partway up the wall from the floor.

7. _____ is an edge of lumber whose sharp corners have been rounded.

8. _____ is a number given to a material to indicate its resistance to passage of heat.

9. _____ is an additional sash placed on the outside of a window to create dead air space to prevent the loss of heat from the interior in cold weather.

10. _____ is the side of a piece of wood that looks best or the side that is exposed when installed.

11. _____ is unwanted movement of air into an insulation layer of conditioned space.

12. _____ is a material used to prevent the passage of water in the gaseous state.

13. _____ is a narrow strip of thin metal or other material applied to windows and doors to prevent the infiltration of air and moisture.

Name: _____   **Date:** _____

Job Sheet 1: Estimating Drywall

Before starting a remodel, new construction, or repair project, it is necessary to determine the amount of materials that will be needed to complete the assigned task. In most cases, this can be done at a homebuilder's center by simply providing them with the necessary dimensions. However, material estimation can be easily accomplished with a little practice. In many cases, there are free estimating forms and programs on the Internet that can be used to accomplish this. Below is a sample estimating form in which the amount of drywall can be estimated.

Areas	Height	Width	Height × Width	Total Area
Wall				
Ceiling				
Excluded Areas				
		Base	(Base/2) · Height	
Sloping Wall				

Sheet Rock Size	4 × 8	4 × 9	4 × 10	4 × 12
Area	24	36	40	48

Total Dry Wall Area   = _____
([Wall + Ceiling] − Excluded Areas + Sloping Walls)

Total Drywall Required   = _____
([Total Dry Wall Area]/Sheet Rock Area)

Instructor's Response

Name: _____   Date: _____

Job Sheet 2: Using the Drywall Estimating Form

Using the drywall estimating form provided below, determine the amount of drywall need to cover a wall 10 feet × 12 feet and a ceiling 10 feet × 15 feet.

Areas	Height	Width	Height × Width	Total Area
Wall				
Ceiling				
Excluded Areas				
		Base	(Base/2) · Height	
Sloping Wall				

Sheet Rock Size	4 × 8	4 × 9	4 × 10	4 × 12
Area	24	36	40	48

Total Dry Wall Area   = _____
([Wall + Ceiling] − Excluded Areas + Sloping Walls)

Total Drywall Required   = _____
([Total Dry Wall Area]/Sheet Rock Area)

Instructor's Response

Chapter 15 Interior Finish

OBJECTIVES

Upon completion of this chapter, you will be able to:

Knowledge-Based

- Identify the components of a suspended ceiling system.
- Identify standard interior moldings and describe their use.

Skill-Based

- Apply ceiling and wall molding.
- Apply interior door casings, baseboard, base cap, and base shoe.
- Install window trim, including stools, aprons, jamb extensions, and casings.
- Apply strip and plank flooring.
- Estimate quantities of the parts in a suspended ceiling system.
- Estimate the quantities of molding needed for windows, doors, ceiling, and base.
- Estimate wood flooring required for various installations.
- Lay out and install suspended ceilings.

Keywords

Apron	Spline	Stool
Compound Miter		

Introduction

Once the interior walls have been framed and covered and the electrical and plumbing fixtures installed, then finish carpentry work can begin. This includes installing the flooring, trim, and molding. While this is still a part of carpentry, it requires a different set of skills from framing carpentry.

Estimating Suspended Acoustical Ceilings

The ceiling of choice for most commercial construction projects is a suspended acoustical ceiling. The amount of ceiling tiles needed for these types of ceiling can be determined by simply multiplying the length of the room to be covered by the width of the room and then dividing that by the area of the ceiling tile to be used. For example, calculate the number of ceiling tiles that will be needed for a room that is 10 feet × 15 feet, using 2 foot × 4 foot ceiling tiles.

1. Calculate the total area of the ceiling to be covered.

   Ceiling Area = length of room × width of room
   Ceiling Area = 10 ft. × 15 ft.
   Ceiling Area = 150 sq. ft.

2. Calculate the total area of the ceiling tile.

   Ceiling Tile Area = length of ceiling tile × width of ceiling tile
   Ceiling Tile Area = 2 ft. × 4 ft.
   Ceiling Tile Area = 8 sq. ft.

3. Determine the number of ceiling tiles to use.

   Number of Ceiling Tiles = ceiling area/ceiling tile area
   Number of Ceiling Tiles = 150 sq. ft./8 sq. ft.
   Number of Ceiling Tiles = 18.75 or 19

Constructing the Grid Ceiling System

The following steps outline the procedure for installing a grid ceiling system.

1. Locate the height of the ceiling, marking elevations of the ceiling at the ends of all wall sections.

2. Fasten wall angles around the room with their top edges lined up with the line.

3. Make miter joints on outside corners.

4. From the ceiling sketch, determine the position of the first main runner.

5. Install the cross tee line by measuring out from the short wall, along the stretched main runner line, a distance equal to the width of the border panel.

6. Install hanger lags not more then 4 feet apart and directly over the stretch line.

7. Cut a number of hanger wires by using wire cutters (about 12 inches).

8. Stretch lines, install hanger lags, and attach and bend hanger wires in the same manner at each main runner location.

9. At each main runner location, measure from the wall to the cross tee line and transfer this measurement to the main runner.

10. Cut the main runners about ⅛ inch less to allow for the thickness of the wall bracket.

11. Hang the main runners by resting the cut end on the wall angle and inserting suspension wires in the appropriate holes in the top of the main runner.

12. The length of the last section is measured from the end of the last one installed to the opposite wall, allowing about ⅛ inch less to fit.

13. Cross tees are installed by inserting the tabs on the ends into the slots in the main runners.

14. Lay in a few full-size ceiling panels to stabilize the grid while installing the border cross tees.

15. Cut and install cross tees along the border.

16. For 2 × 2 panels, install 2-foot cross tees at the midpoints of the 4-foot cross tees.

17. Ceiling panels are placed in position by tilting them slightly, lifting them above the grid, and letting them fall into place.

18. When a column is near the center of a ceiling panel, cut the panel at the midpoint of the column.

Chapter Review Questions and Exercises

COMPLETION

1. _____ is a piece of the window trim used under the stool.

2. _____ is the bottom horizontal member of interior window trim that serves as the finished window sill.

3. _____ is a thin, flat strip of wood inserted into the grooved edges of adjoining pieces.

4. _____ is a bevel cut across the width and also through the thickness of a piece.

Name: _____   Date: _____

Job Sheet 1: Calculating Ceiling Tiles

- Upon completion of this job sheet, you will be able to calculate the number of ceiling tiles needed for a particular application.
- Calculate the number of ceiling tiles that will be needed for the following measurements.

Room			Ceiling Tiles			
Length (Ft. In.)	Width (Ft. In.)	Area	Length (Ft. In.)	Width (Ft. In.)	Area	Number of Tiles Needed
8 ft. 4 in.	8 ft. 9 in.		2 ft.	4 ft.		
9 ft. 9 in.	10 ft. 4 in.		2 ft.	2 ft.		
13 ft. 9 in.	9 ft. 4 in.		2 ft.	2 ft.		
11 ft. 5 in.	11 ft. 5 in.		2 ft.	2 ft.		
12 ft. 3 in.	12 ft. 4 in.		2 ft.	4 ft.		
16 ft. 9 in.	16 ft. 6 in.		2 ft.	4 ft.		
14 ft. 9 in.	14 ft. 2 in.		2 ft.	4 ft.		
15 ft. 2 in.	15 ft. 4 in.		2 ft.	4 ft.		
10 ft. 9 in.	13 ft. 1 in.		2 ft.	4 ft.		

Instructor's Response

Chapter 16 — Stair Framing and Finish

Upon completion of this chapter, you will be able to:

Knowledge-Based

- ◉ Name various stair finish parts and describe their location and function.
- ◉ Describe several stairway designs.
- ◉ Define terms used in stair framing.
- ◉ Determine the unit rise and unit run of a stairway given the total rise.
- ◉ Determine the length of a stairwell.

Skill-Based

- ◉ Lay out a stair carriage and frame a straight stairway.
- ◉ Lay out and frame a stairway with a landing.
- ◉ Lay out, dado, and assemble a housed-stringer staircase.
- ◉ Apply finish to the stair body of open and closed staircases.
- ◉ Install a post-to-post balustrade from floor to balcony on the open end of a staircase.

Keywords

Baluster	Newel Post	Shank Hole
Balustrade	Rake	Stairwell
Handrail		

Introduction

In a multistory structure, stairs are a vital part of getting from one floor to another. In a residential setting, it is the primary means of getting from floor to floor; in a commercial setting, elevators are the primary means. However, in commercial as well as residential structures, stairs are also considered a means of egress (escape) in the case of a fire. Regardless of the application, if the stairs are not designed properly, then not only can they be considered unsafe, but they can also be difficult to navigate.

Chapter Review Questions and Exercises

COMPLETION

1. _____ is a railing on a stair intended to be grasped by the hand to serve as a support and guard.

2. _____ is the sloping portion of trim, such as on grade ends of a building or stair.

3. _____ is a vertical member of a stair rail, usually decorative and spaced closely together.

4. _____ is the entire stair rail assembly, including handrail, balusters, and posts.

5. _____ is an upright post supporting the handrail in a flight of stairs.

6. _____ is an opening in the floor for climbing or descending stairs or the space of a structure where the stairs are located.

7. _____ is a hole drilled for the thicker portion of a wood screw.

Name: _____   **Date:** _____

Job Sheet 1: Identifying the Different Parts of a Set of Stairs

- Upon completion of this job sheet, you will be able to identify the parts of a set of stairs.

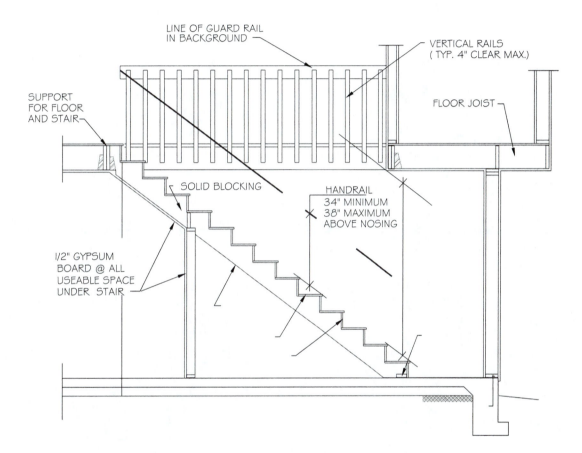

LINE OF GUARD RAIL
IN BACKGROUND

VERTICAL RAILS
(TYP. 4" CLEAR MAX.)

SUPPORT
FOR FLOOR
AND STAIR

FLOOR JOIST

SOLID BLOCKING

HANDRAIL
34" MINIMUM
38" MAXIMUM
ABOVE NOSING

1/2" GYPSUM
BOARD @ ALL
USEABLE SPACE
UNDER STAIR

Instructor's Response

Name: _____   Date: _____

Job Sheet 2: Determine the Rise and Run of a Set of Stairs

- Upon completion of this job sheet, you will be able to determine the rise and run of a set of stairs.
- To determine the rise and run of a set of stairs, use the following steps:

    1. Total Rise = floor to ceiling height + floor joist + depth of floor covering

    2. Number of Risers = divide total rise by maximum rise (7¾″)

    3. Number of Treads = number of risers − 1

    4. Total Run = tread width × number of treads

Determine the rise and run for a set of stairs connecting two floors if the distance between the floors is 9 feet. The second floor is framed by using 2 × 10s and has a ²³⁄₃₂ subfloor with a ⅝-inch wood floor over that.

Instructor's Response

Name: _____   **Date:** _____

Job Sheet 3: Stair Terminology

- Upon completion of this job sheet, you will be able to describe some of the parts of a set of stairs.
- Define the following.

Stringer or Stair Jack

Kick Block or Kicker

Headroom

Handrail

Guardrail

Instructor's Response

Chapter 17 Cabinet and Countertops

Cabinet and Countertops

OBJECTIVES

Upon completion of this chapter, you will be able to:

Knowledge-Based

- ○ State the sizes and describe the construction of typical base and wall kitchen cabinet units.
- ○ Identify cabinet doors and drawers according to the type of construction and method of installation.
- ○ Identify overlay, lipped, and flush cabinet doors and proper drawer construction.

Skill-Based

- ○ Apply cabinet hinges, pulls, and door catches.
- ○ Install manufactured kitchen cabinets.
- ○ Construct, laminate, and install a countertop.

Keywords

Face Frame	J-Roller	Post Forming
Gain	Pilot	

Introduction

In residential construction, cabinets are a major expense item that not only serve as storage, but reflect the individual style of the homeowner. They should be designed and constructed to be both functional and decorative.

Replace and Repair Cabinets

The following steps outline the procedure for installing manufactured cabinets.

1. Measure 34½ inches up the wall.

2. Next, mark the stud locations of the framed wall.

3. A cabinet lift may be used to hold the cabinets in position for fastening to the wall.

4. Install upper corner cabinets first, and then proceed with adjacent cabinets.

5. Align the adjoining stiles so their faces are flush with each other.

6. Install base corner cabinets first, and then proceed with adjacent cabinets.

7. Cut both ends and the toeboard to the scribed lines.

8. After the base units are fastened in position, cut the countertop to length.

9. Fasten the countertop to the base cabinets with screws up through the triangular blocks usually installed in the top corners of base units.

10. Exposed cut ends of postformed countertops are covered by specially shaped pieces of plastic laminate.

Chapter Review Questions and Exercises

COMPLETION

1. _____ is a 3-inch-wide rubber roller used to apply pressure over the surface of contact cement-bounded plastic laminates.

2. _____ is the method used to bend plastic laminate to small radii.

3. _____ is a point of rotation.

4. _____ is a framework of narrow pieces on the front of a cabinet making the door and drawer opening.

5. _____ is a guide on the end of edge-forming route bits used to control the amount of cut.

6. _____ is a cutout made in a piece to receive another piece, such as a cutout for a butt hinge.

Name: _____   **Date:** _____

Job Sheet 1: Kitchen Cabinet Layout Identification

- Upon completion of this job sheet, you will be able to identify various components of kitchen cabinets.
- Identify the items with missing labels.

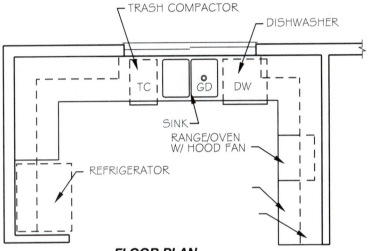

FLOOR PLAN

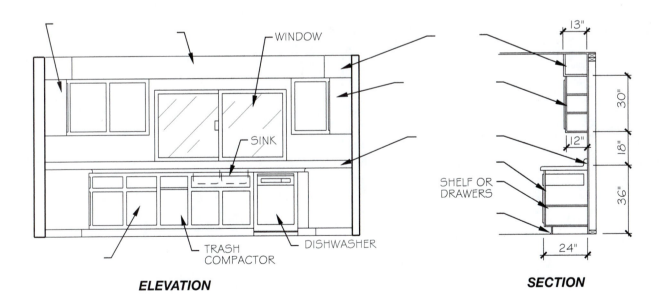

ELEVATION

SECTION

Instructor's Response